DANGER
CONSTRUCTION ZONE

Your **GUIDED TOUR** **TO A** *Successful*
REMODELING
PROJECT

By Steven Katkowsky

AuthorHouse™
1663 Liberty Drive, Suite 200
Bloomington, IN 47403
www.authorhouse.com
Phone: 1-800-839-8640

AuthorHouse™ UK Ltd.
500 Avebury Boulevard
Central Milton Keynes, MK9 2BE
www.authorhouse.co.uk
Phone: 08001974150

First published by AuthorHouse   2/16/2006

ISBN: 1-4259-2006-3 (sc)

Library of Congress Control Number: 2006901423

Printed in the United States of America
Bloomington, Indiana

This book is printed on acid-free paper.

To those at home whom I love the most.
My wife, Jenay, who stayed up nights
and weekends slaving away and
helping me to the bitter end.

To my babies,
who agreed to appear on the back cover
free of charge, Popeye and Pansy.

(Let's not forget my Mommy in Florida.)

Thank you all.
I dedicate this book to the four of you.

Credits
Book cover illustration by Joe Curtis. 831-476-4655

Book design by Marianne Wyllie, Santa Cruz, CA.

Visit the author's website at:
www.stevensincrediblekitchenmachine.com

Acknowledgements

A special thanks to all of my friends who helped me through the ordeal of editing this book. Without their help and input, this would never make the *N.Y. Times* best seller list.

So thank you, Adra Ross, Sheila and Jerry De Lany, Bill and Dorothy Mansfield, Debbie Sanchez, Alice Reed, Norma Lopez, and my brother Eddy Katkowsky.

A special thank you to my friend Joe Curtis for designing the very special book cover.

A Politically Correct Statement

Before printing this book, I had to think of all the different ways I could possibly say the wrong thing. Because there are many women in the trades these days, I had to come up with a way of not saying HE-HIS-THE GUY-HIM-etc. So, how does he/she sound? I didn't think it sounded right either.

There are many women who are contractors, architects, electricians, designers, and salespeople, so I am profoundly sorry if I offend anyone by referring to these professions in the male gender. Doing so simply made the book easier to read.

Table of Contents

Foreword

Think of the first monster movie you saw when you were a child. **WERE YOU AFRAID?** I was. I was mortified. In fact, I slept with the lights on for weeks.

Now imagine those same gut wrenching, frightening, scary feelings coming back to haunt you **IN REAL LIFE,** and maybe, just maybe, you won't even be able to turn on the lights—or the water, stove, refrigerator, toilet, or shower.

These are the obvious things you don't want to have happen. They cause inconvenience. But what about turmoil and divorce? Believe it or not, remodeling your home can cause ALL of the above.

Undertaking a major remodeling project on one of the single largest investments you will ever make, your home, is no easy task. Your home is not only your investment, it is your CASTLE—the kitchen being its HEART, the bathroom its THRONE. This is a project you want to do right. This is a project you want to enjoy.

When going into a remodeling project you should have some long term goals. Ask yourself:

- Will I have the kitchen or bathroom I truly envisioned?
- Will the project be structurally solid, secure and safe?
- Will I have spent an arm and a leg, putting myself into bankruptcy?
- Will I still own my house?
- Will my family still be together and still speaking to each other? More importantly, will I still be married?

Reading this book and following my **FLOOR** plan should help you not only answer these questions, but achieve your goals.

Having been a general contractor and professional kitchen and bath designer for over 34 years, I have learned a great deal about the remodeling process. During the many remodeling projects that I have completed, I have discovered, through successes and mistakes—what works and what doesn't, what pitfalls and problems can arise and how to avoid them.

I've written this book to share my experience and knowledge with you so that you can create the remodeling project of your dreams, and find satisfaction in the process.

So, you're thinking of doing a kitchen or bathroom remodel, or any remodeling for that matter? If the answer to this question is **"YES,"** this book is for you!

By reading and using the information in this book, you will avoid re-living those early, gut wrenching, *scary movie* type emotions.

Consider these pages the best means of **PRACTICING SAFE CONTRACTING.**

About the Author

Please allow me to introduce myself. My name is Steven Katkowsky. I was born in 1948 in Detroit, Michigan—a great place to be from. I now live in California. WOW! What a difference! A great place to be!

I have been in business since I was 15 years old. I remember going to a garage sale and purchasing a set of antique chairs. After refinishing them, I was able to resell them for a profit. This experience was rewarding for two reasons.

1. I made a profit.

2. More importantly, the biggest reward was the smile on my customer's face. **EVEN WHEN THEY WERE WRITING THE CHECK!**

They bought something from me that initially looked scruffy and old. I took the items, did exactly what they expected me to, and presto, a satisfied customer.

This started my early career buying and selling antiques. I did this through high school and college, while acquiring a B.S. (bachelor of science degree) in social work.

Along the way my older brother asked me if I wanted to earn a few extra bucks installing kitchen cabinets. Well, that started a second career in 1970.

I soon gave up the antique business (except of course for collecting), packed up, and moved the kitchen business to San Jose, California.

In 1979 I opened up one of the first large retail kitchen and bath centers in the San Jose area.

I have currently been a general contractor and professional kitchen and bath designer for over 34 years. This business has been very rewarding for both me and my wife, Jenay.

Jenay has been practicing dentistry for over 27 years. She recently retired, and now helps me however and whenever she can.

In 1980, I was asked to present seminars at several home shows.

I attended other seminars and noticed that most of the speakers simply promoted their own businesses and really did not give consumers much information to work with. So I proceeded to put together a presentation that said very little about myself or my business.

My goal was two-fold:

• To frighten consumers into being smart shoppers in this field.

• To frighten all of my competitors (the bad ones) into cleaning up their act, **OR ACT SOMEWHERE ELSE.**

Well, it seemed to work! I have had thousands of people attend the seminars. And occasionally I've seen others in the trades sitting in and listening to the information. I have been approached by grade school teachers, college professors, and most recently a representative from the California's Department of Consumer Affairs commending me for teaching a seminar that helps tremendously in avoiding what can easily become a disaster.

Enough of **ME** for now.

Let's move on to you.

Now it's time for **YOUR** trip. Strap yourself in and enjoy your journey.

I sincerely hope I will be a good tour guide on this adventure.

Introduction

Let's start with the Horror Stories.

These tales seem to abound internationally. I never cease to be amazed at the poor ratings small contractors have all over the world.

When discussing the problems people have encountered, one thing almost always comes to mind.

The only person who can save you is yourself!

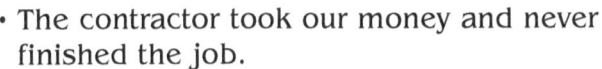

REMEMBER, THIS IS NOT GOING TO BE A REVIVAL MEETING. YOU MUST SAVE YOURSELF.

No one else will!

Now, just what exactly are these stories?

Well, let me list just a few for you:

- The contractor took our money and never finished the job.

- The contractor took six years instead of six months to do the project.

- The contractor didn't have the insurance he said he carried.

- The contractor won't come back to take care of our complaint list.

- The contractor said he was licensed and really wasn't.

- The contractor said that we didn't need permits.

- The contractor's subs were just people he picked up at the local lumberyard parking lot.

- The project ended up costing a lot more than we were led to believe.

- We had some dangerous and unsatisfactory work done on our house.

- We paid our contractor in full. Months later a lien was put on our home by a subcontractor.

The list goes on and on.

HOW COULD THIS HAPPEN?

Hopefully by the end of this book, **your** list will be a **referral** list, **NOT** a **complaint** list.

Throughout this book I will be giving you a systematic way of avoiding **ALL** of the above.

Let me show you:

- How to choose a contractor or designer.

- Who to throw out of your house, and just exactly when to do it, before making a wrong decision.

- How to check out subcontractors.

- How to check out the licenses.

- The correct way to use references.

- How to get proof of insurance.

- How and why to go through the permit process.

- What permits are needed.

- How to avoid winding up with a lien on your property and losing your home.

- How to save your money.

Although I do joke around when covering these subjects, I really need you to pay close attention. Believe it or not, if you follow my system, you will have a very high likelihood of success on your project.

It doesn't take a genius to run a good business. It just takes someone with a good attitude, honesty, skill, experience, a genuine concern for the customer, and most of all, a real desire to be the best at what he or she does.

Chapter 1

The Salesperson
The Designer
The Contractor

Can one person do the job of three?

Can three people do the job of one?

TRICK QUESTIONS? PERHAPS NOT.

I'd like to point out something important before moving on. We will be talking about three different professionals here:

The contractor
The designer
The salesperson

Believe it or not, it is possible to find all three wrapped up in one. It isn't easy, but we are out there.

> Having just one person handling all of the responsibilities for the entire project can be a life saver.

Let me give you an example of how easily a problem can be resolved if you manage to sign up with a person qualified as all three.

Suppose your designer is off on vacation when your new cabinets are delivered. The cabinets come with the wrong doors, and the layout just doesn't seem to be what you agreed on. What are your options here?

- Stop the job until the designer returns from vacation.

- Settle for less than you expected.

- Argue with the cabinet maker until you turn blue and need CPR.

- Talk to your contractor/ designer about what you agreed upon initially and have signed a contract for.

I think you would opt for the last choice. I know I would.

This example addresses several issues.

Examine it for a moment.

- It shows that a major problem can be rectified almost immediately if you have one person in charge.

- It shows that a major problem need not shut down your project for weeks on end if you have one person in charge.

- It shows that you won't have to settle for less than you expect if you have one person in charge.

It clearly demonstrates that an old saying by Harry S. Truman, a past president of the United States (not a contractor) holds true.

"The buck stops here."

No finger pointing, no blaming others, just a simple assumption of responsibility by your experienced contractor/ designer/salesperson—all rolled into one person!

Later on in the book, I am going to summarize a story written in the *San Jose Mercury News* concerning this issue. This story reports on a serious complication during a remodeling project. You will see that if the designer on the project was also an experienced contractor, this disaster could have been avoided.

A word about architects here.

In some instances it is necessary to have an architect involved in your project. If you are making major structural changes to your home, for example, changing roof lines, adding a second story, most anything that changes the aesthetics or structural soundness of your home, this will usually require an architect, and possibly a structural engineer.

It is, in my opinion, a waste of money to hire an architect to simply design your kitchen or bathroom. Use a specialist for this. A kitchen and bath specialist will often be most creative and innovative!

Chapter 2

Let's Get Started

I want you to remember that your home is indeed your castle. Personally, I feel that my home is an extension of **MYSELF.** Never will this become more apparent than when you embark on a remodeling project and hire a contractor. You will now be inviting people you don't know to come into your home and perform surgery on it. Think how carefully you would select a surgeon for YOURSELF. Ever hear the phrase, IT DOESN'T TAKE A BRAIN SURGEON TO FIGURE THAT OUT! – Use it here!

> The people you hire will be one step short of moving in. They will be coming and going during the day, even having a key to your private domain. Your contractor and his subs will become a part of your family for an extended period of time. Your home will cease to be your home, and will become their work area—their *construction site.*

Although I am going to ask you to do a tremendous amount of work researching and investigating these people, don't feel overwhelmed. I cannot emphasize enough how important it is that you play an active role in this selection process. Your diligent and thorough evaluation of the contractor and subcontractors will not only save you time in the long run, but will also greatly reduce the possibility of a SURGICAL MISHAP.

Chapter 3

The Initial Contact

Before setting up your initial appointment with someone, **pay attention.** Look for indications of an irresponsible person or signs of incompetence.

I like to call these signs *LITTLE RED FLAGS.* If you are raising kids, or remember being one yourself, it is not too difficult to detect when someone is not being up front and honest with you.

Here are some key things to look for when SORTING OUT THE GARBAGE, as I so eloquently put it.

Let's start with:

Returning Phone Calls

How many times in your life have you called to set up an appointment with someone and they don't call back in a timely manner,
or never call back at all?

What do most of us do?

Go elsewhere? **NO.**

Call someone else? **NO.**

What most of us do is set ourselves up to be miserable. We wait and wait, hoping for them to call. Or **we** call **them** again.

Well, guess what?

This is the very first indicator *(red flag)* that you do not want to work with this person.

> If a man or woman cannot even make an attempt to impress you on their initial contact, that just may be a hint of what it's going to be like working with them for the entire project.

Remember, there are many good people out there in the construction business. Don't be afraid to write one off, even a good referral. Take the GARBAGE out early—THE TRUCK IS BIG.

Appointment Promptness!

Boy, are we all misguided on this one. I know that in my life I have been kept waiting on numerous occasions. For example, let's say an appointment is set for 6:00 P.M. The person calls at 6:20 P.M. to say they will be five minutes late. Then, another call at close to 7:00 P.M. to say they are stuck in traffic.

Well, some of you might think this very considerate, keeping in constant touch. Then, no one shows!

Ever happen to you? Most likely it has!

Why on earth would you give this person another chance?

WHY?

I have heard many different reasons from customers why they do.

- He sounded so nice.
- His car broke down.
- He was rushing his wife to the hospital.
- blah, blah, blah.

Don't Fall for It!

First of all, this is the twenty first century. Most of us have cell phones. If a person is going to be late for an appointment they should call BEFORE they were due, NOT twenty minutes after the set time. A call letting you know they will be late is indeed considerate, but not **after the fact.**

I once had an employee call in after he was four hours late for work. This happened years ago, but I have never forgotten his excuse.

In a quivering, sad, quiet voice, he told me his grandmother had passed away the night before. I really did feel bad for him, wouldn't you? But I felt bad for a different reason. I knew his parents, and I wasn't aware of them ever having been divorced. I told him this—and made him acutely aware that this being the situation, he would only have two grandmothers.

THIS WAS THE THIRD TIME HE CALLED IN WITH THE DEATH OF A GRANDMOTHER.

He did not get another chance to show up late!

> When someone gives you a poor excuse for being late, let them be late someplace else.

These poor excuses may be indicative of just how the project may go if you use this lame brain.

SOMETIMES A LAME BRAINER IS A NO BRAINER.

Now, you finally have someone coming over who was prompt in returning your phone calls, and you have set up an appointment.

Surprisingly, after what I have told you previously, very surprisingly, they have shown up on time.

AAAHHHHH! What do I do now?

I have a salesman in my house!

What you do next is continue to use your head.

This is where you begin to really qualify someone to the best of your ability. As time goes on, the GARBAGE CAN will become less full.

Remember, people like me are going to try to sell you not just a remodeling job, but sometimes the BROOKLYN BRIDGE may come along in the deal.

Brooklyn Bridge

Chapter 4

Qualifying the Contractor

One thing I cannot stress enough is the importance of using common sense.

Don't be afraid to ask questions. When you are interviewing someone to work in your home, believe me, you need everything short of a **BLOOD TEST.**

These people are going to be working in and on one of the single largest investments you will make in your life, YOUR HOME. They will become part of the family over a several month period, and you don't need the extra worry of babysitting another child—especially a grown up one.

So, here is my roadmap showing you how to know if a contractor is telling you the truth, and how to further check his credentials after he is gone. Without him ever knowing!

Ask questions!

Start here:

- Experience.
- Where the experience came from?
- How many years?
- Where employed in the past?
- References.
- Portfolio, albums.
- Product knowledge.

- Are you licensed?

- Are you insured?

- Do I need to get permits?

- Anything else you feel is important!

The Importance of Experience

Remember I told you that I've been doing this for over thirty years?

Pretty impressive if you ask me. But more than likely, you didn't ask me...

I TOLD YOU SO!

After hearing that someone has so many years of experience, let's see just how good this experience is. Where did the experience come from? Textbooks? Or in the field packing tools of the trade? Practical experience can be crucial.

When hiring a designer or contractor, make sure that he has many years of real, practical experience. This can be checked through references. Also in many states the lower the license number, the longer the license has been active.

The importance of this was recently written about in the *San Jose Mercury News*, San Jose, California. The story was in the Food and Wine Section on Wed. December 4, 2004.

During this particular remodel, the project's designer thought some existing interior walls should be eliminated, and drew up the plans accordingly. When the workers tore out the wall, the roof started to cave in. **THEY WERE LOAD BEARING WALLS!**

A designer with construction experience would have gone in the attic and crawlspace ahead of time to verify this before sending a crew out to do the work.

The entire mess wound up costing the customer **tens of thousands of dollars more than expected,** and over a year to finally finish.

DO YOU WANT THIS TO HAPPEN TO YOU?

This is a pretty good example of why many years of in-the-field construction experience is very important—for both the designer and the contractor.

A person's job stability can be the CORNER-STONE for BUILDING your confidence. If anyone has been working at the same well established business for five or more years, they more than likely will have picked up some good habits.

But, many contractors, designers, and salespeople are very proud to brag about how many places they have been employed.

Would you be proud to tell your closest friends, let alone someone you just met, that you **can't hold a job?**

If people tell you how great they are, and that they have worked at seven or eight outlets in the past 10 or 15 years, to me that spells irresponsibility.

If you have the nerve to ask why they couldn't stay at one place for more than two years, they will usually respond:

- "Oh, the owner didn't know how to run a business."

- "I had all of these great marketing ideas and they were too stubborn to use them."

- And the most common of all—"they didn't appreciate how much money I made for them. They hardly ever came into work, I did everything!"

Once again,
> Use your head
> Use common sense

The kinds of responses above are more good indicators of **(RED FLAGS)** who to dump in the TRASH BIN.

References Are Critical

References, References, References.

The magic, hidden research department of our business. You must ask for references.

But remember, the contractor is trying to sell you a job. There is a reason that the references that he gives you will all seem to be excellent.

Do you think that you are willingly going to be given a bad reference? Not on your life—**UNLESS YOU ASK!**

Contractors are not stupid. Most of us will give you the best three or four jobs we have done, and will make sure they are showroom quality at that. These are **NOT** the only ones you want to see.

How about the job that took too long? Or the job where a mistake was made?

WHOOPS! Did I say mistakes?

Contractors don't do that, do they?

Of course we do. It shouldn't be so important to you that we made one, but more importantly, **how was it handled?**

The usual finger pointing at the *real culprit?*

- "It wasn't my fault."

- "So and so is a jerk"

OR, did the contractor take full responsibility, without you having to go through any group family counseling?

You didn't hire someone to complain about *so and so the jerk.* You hired him to handle the job in a way that is least intrusive to your family, and your central nervous system.

Let me give you an excellent example of a contractor making a nearly disastrous mistake in someone's home. And, it really wasn't his fault.

(Sound familiar? Wasn't his fault?)

Well, that contractor was none other than **me!** I was working on a kitchen in San Jose, California several years ago. Midway through the job, the customers said that they were going out of town for a week and asked if we could re-pipe the house. No problem, they would avoid living through the mess.

Well, they left on Thursday and the plumber started the re-pipe on Friday. He turned off the water into the house and cut out much of the old iron pipe. Because the customers were not coming back for a week, he simply locked up the house for the weekend and planned to come back Monday to finish.

On Monday, we were not the only ones to return. It seems that the customers' trip plans changed over the weekend, and they returned on Sunday. When they discovered the water was turned off, the customer turned it back on. Why bother Steve on the weekend to do such a simple task? Well, they found out why they should have called Steve on the weekend to do such a simple task. It was a mess! Water all over. The living room and dining room were flooded before the customer could turn the water off.

How did I handle this? I could have told the customers that it was their fault for returning to their home early. **Their home.** Not a good idea I thought.

Perhaps it was the plumber's fault? I could have told them that. But then again, I was the one who told the plumber "not to worry," the homeowner would not be returning for a week. Not a good idea I thought.

Better yet, my first response was "let's get this taken care of immediately." The wife asked if she should call her insurance company. I said "of course not." This was my accident, not hers. Within hours, the carpets were removed, the pad was being replaced, and by the end of the day the carpets were reinstalled, stretched and the dryers were running.

Am I a hero? No. I simply took care of the problem—my job!

> Make sure, by asking his references, that the contractor took full responsibility for any mistakes or problems that occurred on the project, and handled them in a swift and efficient manner.

Albums and portfolios are nice. We all have them. But pictures are not always **WORTH THE THOUSAND WORDS** we have been told to believe they are.

You must go to the reference's home to see for yourself.

You must physically go and look, inspect, and scrutinize every aspect of a project. For instance, are the miters in the crown molding tight fitting? Are the granite seams smooth and even? Do all the doors and drawers align correctly? Inspect all the finish details that make one's work stand out above others.

At this time, let me give you a list of questions to ask the references. Don't be afraid to add your own.

- Did the workers show up on time?

- Were the workers pleasant?

- Was the job site cleaned up each day?

- Was debris piled neatly and removed in a timely manner?

- Was the contractor agreeable to changes during the project, or was he argumentative?

- Were inspections passed the first time the inspector came out, or were there multiple correction notices written?

- Did anyone drink alcohol during work or at lunch?

- Did any workers make you uncomfortable when they were in your home?

- If the project did not get completed on time, was there due cause?

These questions, without realizing it, will become your lie detection device. More on that later. I certainly hope you are looking forward to later.

Product Knowledge

When you are talking to a salesperson, designer, or contractor, pay close attention to his body language when he is explaining things to you, or answering your questions.

Most importantly, **does he only promote what he sells,** as opposed to what you know is available? He may be working towards that BRIDGE I mentioned earlier.
THE ONE LOCATED IN BROOKLYN.

A person trying to sell you a kitchen, bathroom or any other remodeling project, should know something about all of the products available for that project.

He does not however, have to
KNOW IT ALL!

Most of us try to keep up on a broad spectrum of products such as:

    Appliances
    Plumbing fixtures
    Flooring
    Countertop materials
    Hardware
    Woods
    Door styles
    Lighting

The list goes on and on.

> There is no possible way for any of us to know everything about all of these products. A good contractor should know enough to generally inform you as to what is available in all these areas, guide you towards what may be best for your needs, and finally, recommend a professional in each area to help you with the final decisions.

Nobody likes a **KNOW IT ALL.**

> But remember, the contractor should advise you of the pros and cons of each item available, not simply steer you towards what he sells or gets a commission on.

Sensitivity

Understanding what you want!

> Make sure that whoever is designing and
> working on your project understands just
> exactly what you want. The last thing you want
> is a person working on your home who is
> selectively deaf!

O.K.
I am going to throw a phrase at you...

"You don't want that! Here, let me show you what you want!"

HAVE YOU EVER HEARD THIS ONE?

If this were one of my seminars, hands would be up all over
the place.

Now as a husband, I can tell you that I have heard that from
my wife for 35 years, and usually she is right. (boy was that
hard to say). But from a salesman who has only known you
for 35 minutes, now that is questionable, rude and conde-
scending to say the least.

Countless times I have sat down with prospective clients
and asked—"What kind of stove do you want, gas or electric?" BAM, two instant answers from each of the two
clients. Two distinct and definite, seemingly discussed,
answers.

But—one answer is gas, and one answer is electric.

Then they both look at each other, trying to maintain com-
posure, and both say at the same time, "but I thought..."

Well, here is what I think. If you, the couple, have no idea what you want at this stage after obviously discussing it, how the heck is someone you have never met before going to know *what you want?*

My classic example of a salesman who thinks he knows what you want more than you do, concerns purchasing a new truck for my company a few years ago. I will try to make this quick, but I am sure you will relate to it in some way.

This *PRO* I refer to as *DEAF.*

To avoid the usual turmoil of dealing with a car salesman, I went to the dealership when they were closed, chose the vehicle I wanted, and wrote down the dealer stock number. I returned the next day hoping to make this an easy purchase. All the salesperson had to do was **LISTEN TO ME.**

When I arrived, I felt truly blessed. Not a salesperson in sight. I would personally get to choose one. Well, I pulled in, stopped, looked down to shift into park, opened my door, and "bloink" my door just missed hitting one of a few vultures which must have fallen out of the sky in the few seconds it took to park.

I took a deep breath and prepared to make this the easiest sale this guy would ever have. His name was Max. A jolly kind of guy wearing no hearing device. A good start!

"Max, my name is Steven. I am here to buy a truck, a **RED** truck. It has to be **RED,** with this list of options. Why don't you bring it up, forget about introducing me to your three sales managers, and I will buy it." I went one step further to make this guy have a great start to his day. I gave him the stock number.

Now, wouldn't you think that I was a dream customer? I not only knew what I wanted, but knew the stock number!

Well, Max didn't even inhale. He was excited and knew just what **HE WANTED** to sell me. He blurted out, "Steve, I can

see you are probably a busy man. Let me show you this white one over here." **WHITE!**
Perhaps I didn't make myself clear enough.

Well, I now did what I have been preaching not to do! I gave Max a second chance. Why? Because I have bought cars before and knew it would be the same everywhere. (Not so with contractors.) I handed him a business card with the stock number and kindly asked him to "Bring it on up, I would like to buy it."

Response—"No problem." Max was so kind and attentive. He proceeded to bring out and introduce me to:

> His credit manager.
> His fleet manager.
> His sales manager.

What did I do?

What should I do?

I thanked Max for nothing and told him and his crew I was leaving. He asked "why?"

I told him because he was **DEAF**, and I hoped he could explain **WHY** to his managers.

Better yet, why don't we title that one,

DEAF OF A SALESMAN

I turned around, got in my vehicle and left. At home I called another local dealer and explained to him the experience that I had with that salesman.

He dealer-traded the truck (that same one) without ever meeting me or getting a deposit, called me the next day to come see it, and gave me a great deal. **HE LISTENED!**

Should I say it again, **HE LISTENED!**

More things you don't want to hear and why you don't want to.

How about this one:

"YOU'D BETTER BUY NOW.
 THE PRICE IS GOING UP TOMORROW."

THIS IS NOT ONLY A CROCK, IT'S THE WHOLE BARREL.

This single sentence is one of the best **RED** flags out there. Most of the time, you will hear it when the salesman calls after the initial visit.

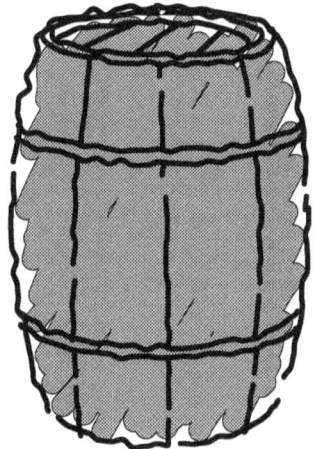

DON'T GET DUPED! OR
      WHOOPDIEDUPED.

Go through your memory banks and try to recall the last big ticket item you purchased other than your house.

REMEMBER?

Remember when you got it home, after shopping for months, city to city, store to store, scanning the internet, taking time off work to ensure the best possible price?

REMEMBER?

What most of us forget for some reason is that the day after we made the purchase, it was on sale for less someplace else.

We are all old enough to know that probably nothing on sale today will not be available soon thereafter for the same or less.

Do you really think that if you are spending fifty thousand dollars or more on a remodel, a salesman is going to turn you away a day or two later—even up to a month later? If he does, he better read the third of this series of books that I have not yet written. It will save him huge psychiatric bills.

Follow-up phone calls are important for both you, the salesperson, the designer, and the contractor. But not to pressure you to buy.

Let me give you just one example of a good follow up call.

When I go to a home to develop a design for a kitchen or bathroom, there is just too much information to cover in one appointment.

Because most people are a little defensive against in-house salespeople, we have to be sure we cover what are the most important subjects. But because countless things are discussed, and everyone is politely interrupting each other with more questions or answers, some information goes by without being fully talked about.

After I leave somebody's home and get into the truck, I immediately make a list of everything I can remember forgetting to ask,or forgot to mention, or simply forgot.

This isn't always so easy!

There were electrical, appliances, flooring, counter surfaces, backsplash materials, fixtures, hardware, lighting, skylights, windows, etc. discussed.

I wait a few days and make my call in the early evening, usually after dinner when I know both people are home.

"Hello, this is Steven. Remember me, the kitchen guy who spent three hours with you the other day?"

"Yeah, what do you want?" Ninety percent of the time, a call not wanted on your end.

I ignore the attitude, and ask if there were any more questions I could answer, or if they need me to come out again.

Usually

A QUICK "NO!"
"NOT NOW!"
"WE WILL CALL YOU!"

Now, I casually mention something which I know that they both wanted to hear and we didn't get to cover. *(This is fairly easy for me. Early on in my dating days I grew accustomed to rejection. I handle it quite well.)*

Then, most people say "Oh yeah, tell me about that briefly, can you?"

This initiates a ten minute conversation, sometimes with both husband and wife on the phone at the same time. Now, the three of us are more assured that most everything was discussed.

A simple thank you for your time, and an offer for them to call or e-mail anything else that may help in their decision-making process is all you need to hear.

Not **"YOU BETTER BUY NOW, THE PRICE IS GOING UP TOMORROW."**

If you don't see this flag waving, I can recommend a good
OPHTHALMOLOGIST.

Checking Licenses

The number **ONE,**

NUMERO UNO!

 # 1 (GOT THE MESSAGE?)
Question to ask a contractor when interviewing him, or her, is

"ARE YOU LICENSED?"

"Of course I am. I told you that I've been doing this for 30 years."

Let's examine that answer. It sure seems to look good. He did come referred. Has indeed been around for a long time.

Makes sense, doesn't it?

Not really! Many people in the trades have been around a long time. Many are perfectly qualified to be licensed.

That does not mean that they are!

Next you ask,

"MAY I SEE YOUR LICENSE PLEASE?"

Not a lot of you would ask that one, would you? It insinuates to the tiniest degree that he is lying.

TOO BAD!

This is one of the best ways to catch someone who is not really licensed. Don't worry about feeling uncomfortable. You can ask this question tactfully. It is the LAW in many states that a contractor carries it at all times when in the field or when selling a job.

Now, he pulls it out, (THE LICENSE), shows it to you, and you are relieved.

WELL, DON'T BE! By this time you may very well be ready to RELIEVE YOURSELF, but

don't be relieved!

At one time in California an extremely large percentage of licenses produced and checked on were bogus. How about this—over 30%.

Check the name on the license, the type of license, and the expiration date.

WHOA THERE

Someone else's name is on the license. "Oh, it is my uncle Bob's license. I have been working under it for years."

This is one of the guys you want out of your house **RIGHT NOW. PERIOD!**

> There isn't one city in the United States that allows someone to work using another person's license.

ANOTHER PERIOD. EXCLAMATION POINT! NEED I SAY MORE? I WILL ANYWAY.

Do you know that it is still possible to catch the guy in a lie?

While interviewing him, look out the window at his truck. Wow! A new one. He must be doing quite well. And his company name is nicely painted on the vehicle with logos. Very impressive!

BUT, WHERE IS THE CONTRACTOR'S LICENSE NUMBER?

Again, in most states it is required by law that the company's license number be printed on anything that has the company name on it. If nothing else, most contractors are proud to have the license, and it is the first thing that they paint on the vehicle.

I guarantee that if the lettering is nice, a pro did it. A professional sign painter should always ask for the license number. He rarely forgets. If a truck is missing it, my bet is that it doesn't exist!

Scary enough?

I am just starting.

This contractor intends to bid your entire project. He is licensed, and you didn't catch him in a lie. Up to now, you've done a great job!

> Well, sorry to inform you that there are different classifications of licenses. The two that you need to be aware of are the following:
>
> 1 *General contractor* license. In California, this is designated with the letter B in the classification spot on the license.
>
> 2 *Specialty trade* license. In California this is designated with the letter C in the classification spot on the license.

- A general contractor can legally do your entire project, including subcontracting all of the specialty trades. Although he is not specifically licensed in each and every trade, he is legally allowed to do all of them should he so choose.

- A specialty trade, like drywall, plumbing, electrical, painting, etc. are usually licensed in their *specific* trade only. These are the subcontractors that the contractor uses on your project. In most states the subcontractors are not allowed to hire out other subcontractors.

Now, everything checks out. The name on the license is correct. It is indeed not a specialty trade only license. It is a B or *general contractor* classification. The expiration date is good. The name and number are both on the truck.

How much better can you do?

A lot better if you try!

Call your contractor state license board and check the license out one last time.

The board will tell you the following:

- If it is indeed a good license.

- If it is current and not expired.

- If the name is correct on it.

- If there has ever been a complaint to the board that has resulted in a citation or legal action.

> • If the license has ever been suspended.
>
> • If the license has ever been revoked.

I would like to comment on the last three items.

Complaint!

Suspension!

Revocation!

COMPLAINT! If there has been a complaint to the State Board by a consumer about their contractor, there is obviously something wrong here.

Think about it. Would you take the time to contact a state agency if your contractor had tried to take care of you? Of course not. Things are usually getting a little bad. If things are going in this direction, Red Flag!

SUSPENSION! Here is the sign of a real loser. If you find out that the license has been suspended FOR ANY REASON, beware! Most states will take much too long to investigate a complaint. This fortunately gives the contractor ample time to rectify a complaint. If he doesn't, he is on his way to the OUTER LIMITS, or living in there already.

REVOCATION OF LICENSE! If a contractor has had his license revoked, ever! **Ever! For any reason! He has to be the most incompetent, unscrupulous, ignorant, brain dead, just short of lobotomized, lazy human being you will ever have the chance to employ.** *(Boy, what a sentence!)*

44

The only thing I would let slide for the moment is an expired license. See when it expired and if it was within the month. Some guys procrastinate until the last minute to renew. **I THINK THIS IS A GUY THING, THE WORLD OVER!** Make sure you follow up on this one item.

Sorry to make you go through all of that, but I just want you to be comfortable with your research project.

I am sure you feel ready to move on.

WELL—SORRY—NOT YET!

A contractor needs more than just a valid license.

Go in the bathroom, look in the mirror and ask yourself:

"AM I A GOOD CARPENTER, ELECTRICIAN, PLUMBER, SHEETROCKER?"

"AM I GOOD ENOUGH TO BE A CONTRACTOR?"

Most likely not.

However, in many states **YOU** can go to a contractors licensing school, take their course and they will guarantee that you will pass the state exam. And then, YOU will become a licensed contractor.

REMEMBER, THIS SCHOOL IS NOT PART OF THE STATE LICENSING BOARD.

It is a private school that makes it easy to get a license. You only need to know how to memorize the questions and answers that they drill into your brain.

These are their qualifications:

- You must have enough money to pay for the class.
- You must show up if you plan to pass the course.
- You must have the minimal electrical activity in your brain to function.
- You must have an advanced degree from at least the third grade.
- You must be able to count to ten.

Of course I have made up these requirements to make a point. I am sure you realize that there aren't this many.

Now that you have attended the school and proven yourself to be a stellar student, there is one more hurdle to leap. The state requires that you have a certain number of years experience working in the trades. How do they verify this? By simply looking at the paperwork that you have persuaded your friends to fill out, verifying the above. Boy that was tough!

Now, would you feel ethical about having your neighbor hire YOU to do a room addition just because you are licensed and have a brand new truck? Don't be fooled by looks. Remember! "Beauty really is only skin deep", and sometimes the SKIN IS PRETTY THIN.

> You must investigate and discover how many years of real experience this licensed contractor actually has. Ask him for references from projects done many years previously.

A few more hours of cross examining your suspect, and he will be vindicated, found not guilty of the above, and put to work in your home.

I am really sorry to make you feel so untrusting—making you feel that you have to be Perry Mason. But, in some cases —,

IF THE GLOVE
DOESN'T FIT,
YOU CAN'T
ACQUIT.

One other point before we continue. Some unscrupulous guys will print their "city business license number" on their business card and try passing it off as a contractor's license number. Make sure you always check with the State Contractor's License Board, or Department of Consumer Affairs, to validate that it is indeed a general contractor's license.

Chapter 5

Don't Forget the Subcontractors

Two more **NUMBER ONE!** important questions. (I GUESS NO ONE LIKES NUMBER TWO ANYWAY!)

When *who is doing the work* is squeezed of all possible answers, you will realize just how important references are.

> Before hiring a contractor, you will need to get a complete list of all of the subcontractors who will be working on your project.
>
> These and no others are the ones whom you will let into your home during the project.

This is very important for a lot of reasons. Again, you are going to use this list in your SHOPPING CART. (A geek phrase. Don't feel bad if you **understand** it.)

If a prospective contractor tells you he has been working in your city for twenty or thirty years, that seems great.

It is even more impressive when he tells you he hires no one out of the newspaper, "my subs have been with me for twenty years."

Tough to do in this business, but entirely possible.

I am really sorry to disappoint you, but you have to be leery here. When you proceed to verify this statement, it could cinch the job for the guy, or send him packing. I HOPE HE CAN AFFORD THE VACATION.

You see, the reason for wanting the list of subcontractors isn't only so that you can check them out.

Ultimately, it will help you come to a verdict in the case of

NOT WHO DONE IT.
BUT WHO WILL DO IT !

What are you going to do with this list of subcontractors now that you have put it together?

First and foremost, make sure that you get the owner of the subcontracted company's name, license number, and business address from your contractor.

Why?

> Because you are going to check the subcontractors out no differently than the main guy doing your project. Look for all of the same credentials.

ALL THE CREDENTIALS!

Everything we have talked about up to now.

Next is where the list of references comes in again. You are going to use the references to **CEMENT** the truth.

> It is **NOT** unreasonable for you to ask the contractor for references from projects older than a year or two. If customers were truly satisfied with the contractor, they will be glad to act as references. This also gives the customer an opportunity to show off their beautifully remodeled home.

Of course you are going to go and see these jobs for yourself to check out the quality of the work. But, just as important is **WHO** did the work?

When looking at a contractor's projects completed just months ago, and another job five years ago, and another ten years ago—casually mention how nice the drywall texture looks and how it matches the rest of the house.

Who did the contractor hire to do the drywall?

Do the same with the hardwood floors, the granite counters, the tile backsplash, the neat and tidy way the plumbing is hooked up under the sink, etc.

If you get four different names out of five different jobs for each trade, **THERE MAY BE A FLAG ON IT'S WAY UP THE POLE.**

This could mean he is hiring tradesmen ad hoc, perhaps even from the want ads. Do you want him doing that? I wouldn't.

Make sure, when you are interviewing the references, that they confirm that the contractor has used the same subcontractors on his projects for many years—never hiring unknowns whose quality of work and ethics are untested.

If the names don't match up to the time period your prospective contractor says he has been using these subcontractors, you may be entering step one of ADOPTING A NEW CHILD TO RAISE!

Most contractors won't even think that you are going to cross examine their clients. Well you aren't. You are simply using them for the reason their names were given to you, as references. For everyone who is involved in the project.

We will jump back to the list of subcontractors occasionally. This is because, as usual, there is more than just one reason we compiled this list.

Chapter 6

Insurance Is a Must

Remember the question "are you licensed?"

I hope so.

IF YOUR MEMORY IS THAT SHORT, MAYBE YOU SHOULD WORK IN THE TRADES !

How about this one?

"Are you insured?"

ANOTHER NO BRAINER, or, a way to find out if you are interviewing a **NO BRAINER.**

"Of course I am insured. How can I be in business with no insurance?"

I have demonstrated how Mr. Unscrupulous can be in business with a lot less than no insurance. You have to follow up on this one as well.

Next, ask "For how much? "

Another tough question, but you must ask!

As you well know if you are a homeowner, you can carry as **MUCH** or as LITTLE insurance as you wish to.

> Would you like someone working on your home with a fifty thousand dollar or seventy five thousand dollar liability policy? In California that won't qualify as a down payment on a lot for a doghouse. You want to hear one million or two million dollar limits.
>
> You want to make sure if there are employees that worker's compensation insurance is carried. Property damage, fire, auto, the whole nine yards.

Well, actually you probably have only one yard so let's rephrase that to the **WHOLE SHEBANG.** (Is that a word?)

Well, now we are on a roll and **I DON'T MEAN EGG ROLL.**

Here is another way to squeeze out just a little more information that may ease the pain down the road.

It is not difficult to obtain this information, and much less harmful than whatever some people **ARE PRESCRIBED FOR MENTAL ANGUISH.**

> I don't want to start sounding like a pessimist here, but don't trust the answers you have been given by the contractor. Instead, have your contractor instruct his insurance company to send you a certificate of insurance naming you as an additional insured for the duration of the job.

Most guys will have their agent send you the certificate showing the following....

- Insured.
- For how much.
- Their address.
- License number.
- Expiration dates.

However, if they skip on their payments and get cancelled **during** the job, you would never know.

> But, if you ask to be named as an **additional insured,** not only the contractor, but you also will get notified of the cancellation.

Pretty darn good **INSURANCE** if you ask me.

WHOOPS

You didn't ask me, did you?

Well, would you like to ask me something? Go ahead. It's time.

WHOOPS AGAIN

GOTCHA

You can't ask me anything! This is a book, not a seminar.
NOW SIT BACK DOWN! Sharpen your pencil, and get ready.
You are on your way to the **NUT HOUSE.** (Better yet, you are on
your way to avoiding it.)

> Insist that all subcontractors do the same with
> their insurance companies prior to start of the
> job. This eliminates unscrupulous or
> irresponsible behavior on anyone's part.

Make sure that if there are employees, that worker's compensation insurance is also listed. (You don't want to be
liable for someone being injured
on your property, do you?)

Remember, having the
subcontractors supply the same
insurance certificates can be a
lifesaver in the end.

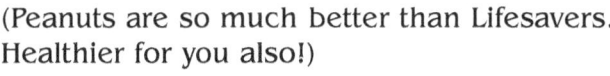

Speaking of lifesavers, let's
move on to **PEANUTS,** I
mean—permits.

(Peanuts are so much better than Lifesavers.
Healthier for you also!)

Chapter 7

Permits

Why?

Permits will prove to be one of the healthiest requests you make in this shopping process, **_TRUST ME !_**

(Boy, I haven't used that line since dating.)

Speaking of dating, I remember some girls that I felt I needed a permit to go out with.

- Don't drive with my baby if it's raining.

- Don't get her home after eleven.

- Don't take her to the drive in.

- Don't get fresh.

- Don't skinny dip.

- Don't get her drunk.

- Don't get her pr#**))^#<>+**# (get it?)

**RULES, RULES, RULES,
REGULATIONS, REGULATIONS, REGULATIONS.**

And, if permits were issued, they were all by her parents.

God forbid if I failed an inspection.

There were never correction notices given—I was simply red tagged.

BANISHED FROM THE JOB FOREVER!

I think you are now aware of the importance of permits, rules, regulations, and laws, especially if you have young daughters.

I will now try to explain the importance of permits relating to another subject, your home remodeling project.

Perhaps—hypothetically of course—you have been wooed and schmoozed over by a slick, smooth talking, impressive Mr. Unscrupulous, and you don't recognize it.

Perhaps you have convinced yourselves that there is really no reason to check this guy out, despite previously reading what I have written up to now.

PERHAPS YOU ARE PLAIN STUPID.

WELL, SALVATION IS HERE.

Hallelujah!

Remember birthday parties when you were a kid?

There was a game that most of us got to play at least once. It was called *bobbing for apples*.

If you miss the importance of the next section, I personally will send you, free of charge, no obligation, especially for you, a mono-grammed, personalized, galva-nized washtub—so you can play **BOBBING FOR BRAINS.**

We are not reading a book here titled **GULLIBLE TRAVELS.** Use your noggin, be smart, be careful, and be shrewd.

PERMITS!

>   Are they necessary?

>   Should you get them?

Here are some *"good"* reasons an unscrupulous contractor will tell you not to get them:

- "Why get permits? Everything is going back exactly the way it was."

- "Permits are so expensive".

- "Permits are so difficult to get."

- "Permits add many days to the job."

- "Get your own permit. Save the money instead of paying me to get them."

- "I'll do the job to code, don't worry."

- "Get a compliance permit when you sell the house later, it costs less."

This is a handful of the **BALONEY** you will be served up by someone who wouldn't have passed most of our investigations up to now.

If a guy can convince you of even one of the above, you deserve him!

Let's go through them one by one:

Why Get Permits? Everything Is Going Back Exactly the Way It Was

Can you explain to me, or your spouse, why you would spend so many $$$$$$$$$ to upgrade your kitchen and put everything back the same way?

In thirty four years I don't think this has ever happened to me in my career.

Here are just a few of the changes made:

- I'd like more counter space.
- Why don't we move the stove or refrigerator to a more convenient location.
- How about moving the dishwasher to the other side of the sink?
- I hate the window. Let's put in a garden window.
- Boy it's dark in here, let's add some lighting.
- Did you know we could put in a skylight dear?
- We are always blowing a fuse when we turn on the microwave and toaster at the same time, let's fix that.
- Sure would be nice if that door didn't swing into the dining room and we could have a pocket door instead. Then, we could move the hutch over.

These are just a handful of the changes people usually make. Make sense?

Permits Are So Expensive

What is expensive? One thousand dollars, three thousand dollars, five thousand dollars, or more?

Believe it or not, if you are not doing a room addition or moving structural walls, all necessary permits, all four...

- Building
- Plumbing
- Electrical
- Mechanical.

...All four—usually come to less than one thousand dollars.

Now that is not expensive considering what it buys you.
I am going to cover that in the next section.

Permits Take So Long to Get

Wow, they sure do. SO LONG I FORGOT TO LAUGH.

> With virtually no experience, and a simple,
> accurate stick drawing, as long as you are not
> making structural changes, you can make an
> appointment in the local building department,
> show up on time, and walk out with your
> permits in less than an hour.

LESS THAN AN HOUR.

I can see now why some guys try to talk you out of it,
or charge thousands of dollars to do it for you.

It may interfere with their golf day, or bar time.

Permits Add Many Days to the Job

How about **THAT** one?

"Permits add so much time to the project due to waiting for
the inspectors. Inspectors are such jerks!" *(per Mr. Unscrupulous)*

Let me explain something to you.

Inspectors are not jerks.

Inspectors are not monsters.

> Inspectors do indeed almost always show up on
> time. I'll bet you don't know that an inspector
> will call you the morning of the inspection and
> give you a mere two hour window of time when
> he will be there.

Get Your Own Permit. Save the Money Instead of Paying Me to Get Them

Getting your own permit will indeed save you having to pay your contractor to do it for you. Once again, you have been told the truth, but I'll bet you don't realize how this can work against you!

Saving a few dollars here may come back and bite you in the end. **THE REAR END**

In my opinion, there is one reason, and one reason only, to get permits on your own. If you are doing the work yourself, or hiring individuals on your own to do the work, get your own permits.

This way,

- **YOU** are responsible for having it done right,

- **YOU** are responsible for the liability insurance,

- **YOU** are responsible for the property damage insurance,

- **YOU** are responsible for the worker's compensation insurance,

- **YOU** are responsible for someone falling off a ladder.

ENOUGH *"YOU ARE RESPONSIBLES"* HERE?

Most of the time when a contractor doesn't want to get your permits for you, **HE CAN'T!**

Why?

- He probably isn't licensed and **CANNOT** get permits to work on someone else's home.

- He probably has no insurance.

- He may be working with someone else's license.

- He may be trying to trick you into thinking it isn't worth it to have permits.

If he somehow convinces you not to get permits, the door is opening, with very squeaky hinges, to a very, very DARK ROOM. REMEMBER THE OUTER LIMITS ?

I'll Do the Job to Code, Don't Worry

Here is another example of Mr. Unscrupulous, someone you really don't know, trying to use that old phrase

"Trust me."
        I wouldn't.

Get a Compliance Permit Later When You Sell the House. It Is Cheaper

Although there **IS** such a thing as a compliance permit, it does not always save your BEHIND.

In fact, trying to have a job inspected down the road after admitting you didn't get a permit may be the best road map around to locating your head. The one with the new brain,

UP YOUR———————————BEHIND

Let's say Mr. Unscrupulous didn't get permits and proceeded to do what he thought he could get away with. Now, several years later, you decide to sell your home, and the real estate company wants to see permits for the remodel! UH OH!

You make the trip down to city hall and apply for a compliance permit. No problem, it is usually issued over the counter in less than an hour.

Boy, this is easier than you thought.

You could even schedule an inspection the following day!

That Steve guy sure didn't know what he was talking about.

"I'll have this deal closed up in a matter of days, Honey !"

Tomorrow arrives and the inspector comes out to see if the job complies. He is already not a happy camper knowing that a job this size was done without permits. He finds an electrical problem, and perhaps a plumbing problem. This work was obviously done around the same time as the remodel, something the inspector can easily spot. Now is when he may bring up the permits and inspections. On a bad day, he may become a little put out!

But he is very fair here.

He does not want to hinder the sale of your home by not passing the three or four required inspections.

What are his options?

He chooses the only fair option he can. At no extra charge or penalty, he will re-inspect what he has just found to be unacceptable. If it is repaired correctly according to code,

YOU PASS!

Oh yeah, by the way, did I mention that at his discretion he may want to see everything similar which may be hidden and covered up under the drywall?

THAT'S RIGHT, INSIDE THE FINISHED WALLS!

> If the inspector finds sloppy work that he can easily see, he usually won't pass as acceptable and legal that which he cannot see, **ESPECIALLY PLUMBING AND ELECTRICAL WORK!**
>
> So, tear open the walls of your remodeled home so a proper inspection can be done. Fair enough?

GEE, STEVE MUST HAVE BOBBED FOR HIS BRAIN A LONG TIME AGO, MAYBE HE CAN TEACH ME HOW TO PLAY!

Hopefully I just made my point why permits are cheap insurance.

ARE PERMITS NECESSARY? YES, YES, YES!

Now, we can happily move onto the types of permits that are needed and the reasons why.

Chapter 8

Types of Permits

1. Let's Start with Electrical Permits

Most homes going through a kitchen remodel were built at least thirty or forty years ago.

Go to an antique store and pick up a *Better Homes and Gardens* magazine, or *Life,* or *Look,* or the *Saturday Evening Post.* They are cheap and will give you a glimpse into the past.

Look at the kitchen appliance ads.

Notice anything?

The counter tops are uncluttered except for the usual bread box, and possibly a toaster.

Back in the *olden days* (how depressing) but, back in the *olden days* we didn't have the following:

Cuisanarts.

Cappuccino and Latte machines.

Microwaves.

Under-cabinet lighting.

Under-cabinet televisions.

Under-cabinet stereos.

Under-cabinet computers.

Coffee bean grinders.

Rice cookers.

SHORT MARRIAGES.

Sometimes I walk into a client's home, am escorted into the future kitchen project, and I have to back out of the house and look on the roof for the *neon appliance* store sign. There is no counter space because they have one of each kind of appliance, mostly plugged into two outlets, on a plug strip that accepts eight plugs each.

I can't believe their first question.

"Is there something wrong with our house? We keep blowing fuses?"

Of course there is something wrong.

> This is a great reason to get permits! Homes were built back when these small appliances were not readily available to everyone. There were usually only two circuits put into the entire kitchen.
>
> • The first one for the dishwasher and disposal.
>
> • The second one for the wall plugs in the kitchen. Almost always, this second circuit also supplied power to other rooms in the house. This is no longer legal, or safe.
>
> The reason the fuses are blowing is to prevent these people from burning down their house. Certain size wires in the wall can carry only so much current. When it gets overloaded, the circuit goes.

Let's look at one example that is usually one of the main culprits.

Most everyone today has a microwave oven. We take it out of the box after getting it home, set it on the counter, and plug it in. HEY GUYS, EVER READ THE DIRECTIONS?

Most of us look at the pictures of how to open the door, stuff something in, push the minute or seconds button, and we are home free. Well, **IF** you looked further, the directions clearly state that this appliance requires a dedicated 15 or sometimes 20 amp. circuit.

That means that this appliance draws at least seventy five percent of the load the circuit is meant to handle when you turn it on.

If you have a plug in the bathroom on the same line as the kitchen, and you plug in a hair dryer and then turn on the toaster at the same time, **BINGO,** you are out in the rain flipping the breaker switch, over and over.

NOW—don't think of flipping your **MIDDLE FINGER SWITCH** at me for making you look stupid, I am simply trying to make a point.

GOOD NEWS!

Having an inspector make sure that the job is wired correctly will save our wives the trouble of having to train us even more than they have already.

WE CAN CONTINUE TO NOT READ THE DIRECTIONS!

Sorry ladies, this is a testosterone thing, we just won't change.

And for you ladies who are taking testosterone, or you guys who have had implants and are on estrogen, the rule still applies.

The electrical codes all over the United States are very similar when remodeling a kitchen. For the most part, they are as follows:

Dedicated circuits for:

Microwave . . . . 15 amp.

Dish washer . . . . 15 amp.

Disposal . . . . 15 amp.

Sub Zero refrigerator . . . . 15 amp.

Plugs. Two 20 amp. Circuits. All of the plugs above counter cannot be on the same circuit.

Instant hot . . . . 15 amp.

Hood/w heat lamps . . . . 15 amp.

Lighting . . . . 15 amp.

Electric oven . . . 30/40/or 50amp.(220 volt)

Electric Burners . . . . 30 or 40 amp.

All of these circuits **MUST** be independent of any others in the house. No more sharing circuits between rooms!

This means that ninety percent of what is currently in the kitchen must be stripped out and redone.

The following is also required:

- Plugs cannot be more than forty-eight inches apart.
- Plugs within six feet of water must be GFCI, or GFI protected. (Ground fault circuit interrupter.)
- You must have fluorescent lighting.
- You must have smoke detectors in the house.

Starting to get the picture?

This is why it is called a remodel.

Electricity is definitely a necessary evil. But does it have to be *evil?*

Not really. Not even if your electrical panel needs to be upgraded. And, guess what? It isn't that expensive.

You're going to get a $SURGE$ of relief here. (get it?)

"My electrical panel is not big enough to handle all of this. It's going to cost a fortune" (says the customer.)

"Hey man, everything seems to work now. How bout we just skip the permits, take a few things off this circuit, and add a few to the ones you don't use so often. **TRUST ME, IT WILL BE O.K.**" (says Mr. Unscrupulous)

Hardly a **point** well taken.

THE ONLY POINT IS THE ONE ON HIS HEAD.

Remember the dating examples I used earlier?

Let's examine his statement for a minute.

"TRUST ME, IT WILL BE O.K."

WHY? Why would you let someone use a line on you like that? A line that you have not used yourself in years?

So, how overwhelming is this problem to solve? Let's just take a look at it:

> A good electrician can strip the wiring out of a kitchen, once it is gutted, in less than a day and a half. A complete rewire, including lighting, might take two and half more days. As for the panel, another one-and-a-half day project. All of this can be done while other work takes place. The total cost will vary state to state, but believe me, this is not a big expense in the overall picture.

Remember, the only charge Mr. Unscrupulous will be saving you is $$$$$$$$.

The charge you will never forget is —

WHEN YOU GET THE BILL FOR REBUILDING YOUR SMOLDERING HOUSE BECAUSE THE INSURANCE COMPANY WOULDN'T PAY FOR THE BONFIRE!

It is the contractor's liability for the fire. (Whoops! He wasn't licensed, or insured. He didn't get permits!)

SORRY

> I am always finding very dangerous things that are hidden in the walls.
>
> · Wires just twisted together and not in electrical boxes.
>
> · Wires with the safety plastic coating melted off due to overload.
>
> · 2X4 framing with wires resting loosely, bare and arcing. Yes, just short of causing a fire.

I can honestly say that in hundreds of homes over the years, we have opened the walls and found sloppy and dangerous electrical work done either by homeowners not knowing better, or creepy *you know whos.*

REMEMBER THEM?

Now, as a homeowner, do you or your spouse feel capable and qualified to do the above?

Hopefully not!

I do hope you now feel qualified to make the intelligent decision to have a licensed, insured, electrician do the work. **WITH PERMITS!**

Not a handy man!

Now, just exactly what will the inspector, the **HORRIBLE CITY MONSTER,** look for on your job? Of course he is going to make sure that **ALL** of the previous electrical list was done, and done correctly.

Think it's that easy? Not much more to look for?

THINK AGAIN!

Here are just a few more things he may be looking for. These are the ones someone not licensed or without the correct experience, may not know about!

> · Are the wires centered on the studs, one and a half inches from the front edge? This helps to prevent the drywall man from putting a nail or screw through it.

- If the wires are too close to the front of a stud, are there steel safety plates nailed onto the stud in front of the wire? This also prevents a wire from being punctured.

- Is the wire stapled every four or so feet so there is not a lot of loose wire in the wall?

- Are there too many wires in any electrical boxes? If you look inside a blue plastic electrical box for example, you will see a rating for how many cubic inches of wire are allowed in it, in various sizes.

- Were safety connectors used in metal boxes to prevent wire from being cut by the box and shorting out?

- Are the breakers installed correctly?

- Is the new electrical panel grounded along with the hot water heater?

In many older homes, the main ground wire goes to the main water lines coming into the house from the street. When the cities change this main water line from iron to plastic,

WHOOSH!
THERE GOES THE GROUND!

A new grounding device is now needed. Does Mr. Unscrupulous know this?

There are many more things that the inspector will check— I cannot possibly mention all of them here. But as you can see, electrical permits are a small price to pay for peace of mind.

Are you starting to realize by now that you really do need a permit?

#2 Well, It's Time to Move on to Plumbing

If burning your house down doesn't scare you, perhaps being able to operate a motor boat in it will.

Or, if interstellar travel interests you, wait until we talk about gas lines.

Plumbing seems too simple, doesn't it?

How soon we forget that every time we try to fix a leak, if we are successful, it only took:

- **ALL DAY.**
- 50 trips to the hardware store.
- The development of a whole new vocabulary.

Plumbing is one of those things that looks so much easier than it is.

Usually the sink and dishwasher are indeed going back to where they were originally. Why would you need a plumbing permit?

Well, it is much cheaper than scuba lessons.

Lets start with your pipes.

Most older homes have iron pipe. Usually, it is noticeable that you are just not getting the water pressure you used to. Although this can sometimes be caused by the fixtures, most times the old piping is simply rotting from within.

It is time to re-pipe the house.

Re-piping a home means replacing all of the iron pipe water lines with copper pipes.

Copper is easy to work with, and a good crew can usually do an average sized home in three or four days. Surprisingly, there is only a nominal amount of wall repair to do, not the **BOMB** effect most of you are expecting.

But, due to the kitchen or bathroom project costing just a little more than you may have expected, a re-pipe may be out of the question for now. You know it needs to be done, and will most likely do it *within a few years*.

Well, check out this fact!

MOST PEOPLE DON'T. They wind up waiting five to ten years!

What happens now, if the plumbing is not done correctly **AND NOT INSPECTED?**

NOTHING! NOTHING HAPPENS NOW! IT HAPPENS LATER, WHEN YOU LEAST EXPECT IT!

Remember the unscrupulous person I keep talking about?

He will give you some good advice here. (Or so it seems.)

"Let's just change the water lines in the walls of the kitchen to copper so you won't have to tear the walls open at a later date," says Mr. Unscrupulous.

Good advice?

So, while the kitchen walls are open, he removes the old iron pipe and replaces it with copper. Now the rest of the house will still be the old iron pipe, something you plan to change later, and only the kitchen will be copper.

Mr. Unscrupulous told you the entire, yes, the **ENTIRE** truth, right down to not having to tear open the walls at a later date!

You won't have to tear them open!

You will simply be able to peel the drywall off like mush.

In fact, some of it will **BE MUSH!**

That is why the house started to smell musty five years later.

You see, you cannot connect two incompatible metals together. There will be a chemical reaction that causes them to slowly rot and again, **BAM!** A new surprise.

WHY TRAVEL TO NIAGARA FALLS?

You can be there any time of the year depending on when the pipes break. And you can be a big shot. Invite the neighbors over, save them money on their next trip.

Of course this is the extreme, but it does happen.

> The inspector will make sure that the copper and the iron pipes are joined by a dielectric fitting. This will prevent electrolysis from causing a flood.

If your house does flood, a good insurance adjuster will look for the cause. Remember, insurance companies don't want to pay if they don't have to.

If the flood was caused by a contractor, and he was licensed and insured,(remember, I covered that?) it will become **HIS** responsibility.

Now, if he wasn't licensed, and he wasn't insured, and it wasn't inspected, and it was illegal, and you knew it...

Now the plumbing inspector is not DRACULA, TRYING TO SUCK BLOOD FROM A TURNIP.

He isn't going to FLY IN, ATTACK AND FLY OUT.

Here are just a few other things he will check on:

- Were the correct types of water supply lines used from the shut off valves to the fixtures? Did you know that the nylon flexible ones are not acceptable in many states? (Call your local building inspector to find out.)

- Are the holes in the floor where the pipes come through filled with foam insulation preventing cold air and rodents from coming up from below?

- If new vents and drains were installed, have the pipes been capped and filled with water to check for leaks?

- Were anti-hammering devices installed at the hot and cold water lines at the sink?

- Are the drain lines from the disposal to the air gap hanging too low?

- Was a new gas line run for the stove?

I am not going to let the gas line slip by us here.

If I do, it is possible to **BLOW** the entire project out of proportion, **LITERALLY!**

For some reason, most people want to put in gas ranges, as though they were going to become gourmet cooks the day the job is finished. This is very good for the appliance companies. Especially the manufacturers that make the shiny, impressive, stainless steel commercial ones.

For those of us who can't afford that Ferrari or Corvette, here is the chance to impress all our friends. **BOTH OF THEM.** You will be able to impress them for a long time. More years than a mere car will last.

Once again, here is where you can be taken advantage of by an unlicensed handyman, who is **NOT ONLY MUCH CHEAPER,**
but also the big **CREEPER.**

"HEY DUDE. No problem to put the gas stove in anyplace you want it. In fact, the gas line from the furnace runs right under the house a few feet from where the new stove is going to be. How's about we just tie into the gas line and bring it up to the new stove? **NO NEED FOR AN EXTRA PERMIT. THE GAS IS RIGHT HERE. JUST A FEW FITTINGS AND ABOUT TEN DOLLARS OF PIPE!"**

Here is another place where the red flag should be flapping in your face!

In fact, right now is when you **SHOULD BE GETTING GAS!** And, maybe consider **SHARING IT WITH THIS NICE PERSON.**

I am going to try explaining this one in simple terms.

To understand this the easiest way, take a trip to Home Depot and go to the plumbing department. Take a look at all the different sizes of pipe. Now ask yourself, "Why do they make so many sizes?"

It's very simple.

Only so much volume (**I DON'T MEAN SOUND**) of anything will travel through a confined space. The different sized pipes will carry different amounts of liquid or gas through them.

Just because you have a gas line under the house, near the kitchen, doesn't mean you can tap into it.

Here is what really should be done.

Find everything that is tied into that line. For example:

- Hot water heater.

- Furnace.

- Gas fireplace

- Gas dryer.

- Pool heater.

- Whatever else.

Add up how many BTUs each one uses. Then, the total lineal feet of pipe from the meter must be measured to these appliances. Measure the size of the pipe, ½", ¾", or 1". There is a formula, to calculate how many BTUs of gas will travel through the existing pipe.

BINGO!

IF THIS WASN'T DONE, YOU HAVE CREATED THE NEXT RESIDENTIAL SPACE SHUTTLE.

If that seems a little complicated, just ask Mr. Unscrupulous to do it for you. I guarantee he will say "I will get back to you."

When you go appliance shopping and decide to purchase that new six oven, twelve burner, three griddle, four broiler, two deep fryer range, you may notice that the gas consumption is similar to heating a six unit apartment building. (just kidding) However, you will see that most of them require a dedicated three quarter inch gas line for the thirty six inch or larger unit. If you simply hook into the existing gas line, turn all the burners on, and then throw something into the oven, you will suck the pilot lights out on everything else.

You may not have another chance to... **WAKE UP AND SMELL THE COFFEE.**

Of course I am exaggerating a little here but I think you get the picture. Someone experienced needs to do this work.

Here is another true, funny, yet sad story of a licensed Mr. Unscrupulous. Yes, some of the licensed ones are also rascals and I don't mean Little Rascals.

I have a friend who is an inspector in one of the local cities I work in. Because it is not a good idea to fraternize with the contractors who he routinely inspects, he would be sure to scrutinize those of us who were his personal friends more closely. He would always make sure that absolutely nothing slipped by—sort of saving his **REAR END,** if you may.

Well, one day he came to one of my jobs to inspect a new gas line I had just installed.

Before it gets hooked up at the meter, the new gas pipe is simply capped at the furthest point in the house. At the meter outside, we attach an air gauge. Then we pump fifteen pounds of air into the pipe and leave it for twenty-four hours. The inspector returns the next day to inspect it, making sure no air has escaped. If any air had escaped, this would indicate a bad connection in the line, requiring it to be repaired.

Usually the pipe comes out of the wall from the crawl space and sticks out about fifteen inches.

Well the inspector, my friend, looked at the meter and said all was well. Then, quite unexpectedly, he reached down, put his foot on the side of the house and pulled on the pipe with all his might. This was something I had never seen done before. He started to giggle and said, "Good thing Steve."

Why did he do this?

Another contractor had called for a similar inspection just a few days earlier. It seems he wasn't quite ready, so he tried to get away with the following:

There was supposed to be about fifty feet of pipe run under the house. Instead, he—

- Took a twenty-four inch piece of the correct sized pipe.

- Capped one end of it.

- Attached the pressure gauge to the other end.

- Drilled a hole in the foundation just big enough for the pipe to fit into.

- Stuffed the pipe into the hole.

- Pumped in fifteen pounds of air.

- Called for his inspection.

Now, this inspector was smart. He tapped on the gauge to make sure it wasn't stuck at fifteen pounds. (Usually they do this because the gauges are cheap and sometimes don't work correctly.) He noticed that the pipe and the gauge moved a little when tapping on it.

It doesn't take a genius to realize that if you have fifty feet of pipe, connected every six feet or so to the floor joists, it will take a bulldozer to move the pipe side to side, not just a simple *tap tap tap.* That is when the inspector grabbed the pipe and pulled it out of the wall.

Now, do you understand why this contractor may be one of those who no longer likes inspectors?

His name flowed like melted butter from city to city, alerting inspectors to be on the watch for this guy's sloppy work!

I think every city in the area gave him heartache for years. Did he deserve it? What do you think?

What if Mr. Unscrupulous had fooled the inspector, ran the pipe later, and had made a mistake?

KABOOOOOOOM!

3 How about a Building Permit?

Why would you need a building permit? We're not adding a room to the house.

- We're just making the window bigger.
- We're just changing the swinging door to the dining room to a pocket door.
- We're just putting a skylight in.
- We're just opening the wall between the dining room and the kitchen.

These are the most common changes made in a kitchen remodel. And, these are just four reasons you need the building permit.

"Why?" You might ask.

Because, each of these changes require you to change the structure of the house.

ANY TIME YOU CHANGE A LOAD-BEARING WALL, IT REQUIRES A BUILDING PERMIT, SOMETIMES ENGINEERING, AND ALWAYS, AN INSPECTION!

How about this one?

The House of Cards

A few years ago I went to bid a job for a *do-it-yourselfer*. When I drove up, the dip in the roof was quite noticeable. It did PEAK my curiosity. Inside I saw a brand new skylight in the area of the new kitchen. The lady of the house was smart. She noticed that after her *genius* put in the new sky-light, the roof started caving in. Well, it seems her husband had cut four trusses to make the hole in the ceiling.

Bad dog, bad dog. Go outside and stay there until you smarten up.

Remember, there is a reason that there is lumber under your drywall. You can't just cut it out and throw it away!

By the time the structural engineer got back to the contractor with what needed to be done for the repairs, a month had passed. **AND IT TOOK SEVERAL THOUSAND DOLLARS TO REBUILD THE HOUSE!**

Here is something to consider.

When selling your home, many states have **DISCLOSURE** laws. You must disclose what you did to the house while you lived there, as well as anything that is wrong with the house.

WITHOUT A BUILDING PERMIT FOR THE REMODEL, YOU JUST MADE A VERY MARKETABLE HOME—

NEGOTIABLE!
>**OR WORSE YET,**
>>**A NO SALE!**

4 Mechanical Permits

A mechanical permit is usually needed for heating and venting ducts when in conjunction with a kitchen or bathroom remodel.

Here, the inspector will make sure that all vent pipes from exhaust fans are indeed vented to the outside, not just into the crawl space or attic.

But is it really this simple?
>Of course not!

I am going to give you just one example here. But it is the most common mistake made by an unlicensed person or an unknowing *do-it-yourselfer.*

When moving your hood in a kitchen, it is usually acceptable to use the existing hole in the roof and existing roof jack. Getting to that area is commonly where the mistake happens. It seems easy to just use the flexible six inch or eight inch expandable venting pipe. It's easy, it's cheap, and it's **dangerous!**

> You cannot use flexible accordion-type pipe to vent a kitchen hood. The creases collect grease over a period of time.

If you ever have a fire on the stove that gets into the hood and up into the pipe, you may be eligible for a **HEAD TRANSPLANT!**

Well, I hope I have made my point about inspections and permits. These guys checking our work are just doing their job. If they catch a legitimate mistake, they will usually let us continue on, and check to see if it is corrected on their next trip out.

They are not jerks!

Inspectors can be true lifesavers.

Point made!

Is it possible that an unreasonable inspector exists out there?

If so, what can be done about it?

Here we go again with one of my stories. But this one is going to turn into a multiple choice test for you.

Remember those? (Better take notes!)

I was recently working on a bathroom—a relatively simple remodel. One of the upgrades was to install a recessed light in the shower. For years the customers felt as though they were standing in a cave.

I called for the rough electrical inspection, set up a time, and the inspector showed up precisely on the hour, just as I have told you he would. He got out of his car, and I couldn't believe how large this guy was. He had to be six foot eight inches tall. He looked like a professional wrestler. He was new—must have been! I certainly would have remembered him if I had met him before.

Well, he came in, introduced himself, and proceeded to do his job. He also proceeded to tell me that he was not going to pass the inspection.

This came somewhat as a surprise. I couldn't find anything wrong. I asked for the explanation and he told me that the light in the shower had to be suitable for wet locations.

Actually, I told him, the code requires the lights in a shower be suitable for **damp** locations. A wet location would be an outside fountain or swimming pool light, usually submerged!

He didn't care what the code said. **HE** was tall enough to stand in the shower and work on the light or change the bulb without a ladder. So, he wanted it changed.

Now, what could I do in this situation?

I couldn't imagine the customer, standing stark naked in the shower with the water running, on a metal ladder, working on a partially disassembled light fixture, changing the bulb. Can you?

How about we take that test now?

What would you do?
Choose one of the following:

    A. Increase the depth of the shower and...

        Install a swimming pool light in the ceiling along with two matching ones at waist level.

        Retro fit the curb with a diving board.

B. Increase the depth of the shower and...

Install an underwater fountain light in the ceiling.

Add two lights in the floor.

Install pneumatic water pumps to shoot the water up like in Las Vegas.

Now when the tour bus from Italy stops by, admission can be charged to throw

THREE COINS IN THE FOUNTAIN.

C. Call a supervisor.

FINALLY, YOU GET 100% ON AN EXAM!

Chapter 9

Liens and
How to Avoid Them

Now that you have become a genius, and have picked *one of the chosen few* to be your contractor, I am going to further destroy your confidence.

Did you know that you are still susceptible to a good tar and feathering?

ARE YOU STARTING TO THINK THAT I'M A CREEP?

AN UNHAPPY,

UNTRUSTING,

PESSIMISTIC SO AND SO?

Too bad. This is why you bought this book, and I am not going to let you forget it until the very end!

The most serious, life threatening shock you can receive as a homeowner, is finding out that **SOMEONE ELSE IS TRYING TO OWN YOUR HOME.**

Someone else whom your contractor did not pay!

That scariest of phrases, **THERE IS A LIEN ON OUR HOME.**

Believe me, again, *trust me,* and here you really can.

You don't want this to happen to you!

If someone puts a lien on your home, it is no longer yours to do with as you please. You cannot borrow money on it, you can't sell it, and you really can't do anything with it until these liens are taken care of.

In some cases, a disgruntled person can foreclose on the lien **AND MOVE YOU OUT!**

How can this happen, even when you have been so careful up to now? I will tell you. Best of all, I will tell you a fool proof way to avoid it. Fool proof!

The only **FOOL** in the end will hopefully be someone else.

Let's say that you were happy with the remodel, loved your contractor and paid him. Actually, you had nothing but a great experience all the way around.

You never asked for the following because I was not smart enough to write this book soon enough. Perhaps, you were too cheap to buy it soon enough.

Unscrupulous contractors are notorious for not paying their subcontractors. Just because you paid your contractor, does not guarantee he paid his subs.

This is where much of what I covered in the opening pages of this tome will come into use, again, for the fiftieth time.

Remember that list of subcontractors I told you to get early on? Now you are going to put the list to work again!

There is no way to really check out if your main contractor is financially solvent enough to pre-pay his subs before he asks you for any money during the job. (He should be!)

So, you should make it clear to your contractor that nothing other than the agreed upon deposit will be paid out without the most important piece of paper in your life being produced, Other than your marriage license, **"AN UNCONDITIONAL WAIVER OF LIEN."**

UNCONDITIONAL!

When you are asked for payment, it should never be because the contractor has to pay so and so. It should always be because he **PAID** so and so!

> Now, when the contractor is politely asking you for money, insist on an **"UNCONDITIONAL WAIVER OF LIEN."** It must be signed by the subcontractor on your list, for work done for your contractor, for you, at your address, marked **PAID IN FULL, ZERO BALANCE.**
>
> There are few business owners who will agree to sign one of these unless he has indeed been paid in full and the check has cleared.

You never, I repeat, never want to pay a contractor with the hope, sometimes dream or fantasy, that he will pay a sub-contractor.

This *"Unconditional Waiver of Lien"* is a two dollar form that you can pick up at any office supply store, in the business form section. Have a few handy in case your contractor says he forgot, or ran out. Have him call his subcontractor on the cell phone to come over and sign it.

If he won't, I wonder why?

Make sure you get one from everyone on your list of sub-contractors. Hopefully, you made sure that no one other than those listed worked on your home.

If any of the subs do change during the job, make it clear to your contractor that you want the same information on them as the original ones. Check them out just as thoroughly, and get waivers of lien from them also.

There is another way to protect yourself, although a little more complicated, but certainly just as effective.

Arrange to pay the subcontractors directly during the project. Have them furnish the waivers, and deduct these payments from the balance. When doing it this way, you usually have to make arrangements for some payments to the contractor for materials and margins during the job.

I personally like the first option better for everyone involved.

Chapter 10

Money $

MONEY, MONEY, MONEY, *HONEY!*

Boy, it seems like life revolves around it. For millennia, money has been the **CORNERSTONE** of society. Make it, spend it, save it, conserve it, preserve it, hide it, stash it, and worship it.

Money, Money, Money.

For most of us, hopefully, money is simply a means to an end. If we are smart, we define those ends and try to live within them.

There is another **END** I have been talking about throughout this book. I don't mean to get too personal, but I hope you realize it by now, I am speaking once again about saving **YOUR REAR END.**

Don't let this next section slip through the **CRACK.**

For some reason, many bad tales in my business revolve around this important means of exchange. It really is not difficult to keep your funds under your control during the remodeling project. Let me show you how.

> Right from the start, the fixed price contract is your best insurance policy for keeping things in control. Making sure that your contract contains everything could save you from having to consume large amounts of nitroglycerin during the project.

Let me cover some of the most common mistakes made by unknowing consumers when signing a contract.

Starting Dates

DON'T ever sign a contract unless a guaranteed starting date is written into it. (you can be flexible within a couple weeks of this date.) If you are continually put off to a later date after signing on, some products that are not included in the job may go up considerably. For example, the cost of appliances and plumbing fixtures. This causes *anxiety*.

Possible Hidden Costs

DON'T ever sign a contract that has open areas where you can be charged for unforeseen, hidden problems. This is called an *"Open Ended Contract"* and can wind up costing you a fortune.

Your contract should be **exact** with a guaranteed **fixed** price.

Another type of contract, rather than the one with the fixed price, is a contract written up where all of the work is done on a *"time and material"* basis. **Bad, Bad idea!**

Many people think that they will save money by running around buying all of the materials and then paying the contractor hourly to do the job.

Some think that by having the contractor buy all of the materials and charging hourly for the job is another way to save money. **Very bad. Very bad idea!**

These are not ways to save you money, they are ways to help the contractor buy that new boat or motorcycle. **HOURLY** can very easily turn into **DAILY!** Be careful here!

This brings to mind the old joke about *how many guys does it take to screw in a light bulb?* Do you want the answer to be **THE ENTIRE CREW WORKING FOR THE WHOLE DAY?**

Remember the words, **FIXED PRICE.**

Now I am going to contradict myself. (I can do this because my wife told me I could.)

There are a FEW unforeseen situations that might arise during a project that may not have been included in the original fixed price contract.

If something like this happens, have your contractor point it out, and again, give you a fixed price on a *"work change order."*

Here is an example of such a situation:

I once tore out a bathroom that had a full tile tub surround. The wall in my opinion was rock solid. However, upon removing the tile, I found the termite damage so catastrophic that there were virtually no studs left in the wall. It was only the cement and tile holding the room together. No Superman— no Kryptonite here to help me see through the wall.

I waited for the customer to come home, explained what had to be done, wrote a *"work change order"* with a fixed price for the additional repairs, and all was well!

What if the situation arises where **YOU** request extra work to be done on your project? For example, crown molding around all of the other rooms in the house. Again, get a fixed price on these extras. No *"time and material."*

> Any time there is to be extra work done that is not on the original fixed price contract, have the extra work written up on a *"work change order."*
>
> This *"work change order"* should be signed by the contractor and you and have the exact price written down!

REMEMBER—FIXED PRICE!

A fixed price usually guarantees that it will take a reasonable amount of time to do the job.

Hidden cost surprises can cause **HIDDEN HEARTBURN.**

Payment Schedules

> **DON'T** sign a contract unless the payment schedule is detailed out for you. Try to keep the deposit reasonable, and make sure all payments during the job are included, in detail, down to the final payment.
>
> And, speaking of final payment, make sure it is a **hefty** one.

Early on I mentioned that one of the horror stories in my business revolved around getting a contractor to return to finish small items.

If you only owe five hundred dollars on a fifty thousand dollar job, a chipped tile may become not *high* on your contractor's priority list. However, if you owe him fifteen thousand dollars, he will probably repair any problems rather quickly.

In addition, owing him this hefty balance is usually a good incentive for the contractor to not only finish the job on time, but to stay in town until the job is indeed finished!

Having this large balance at the end of the job gives you leverage.

I guarantee he will bend over backwards trying to please you here. Don't set yourself up to feel like it is you who is bending over here. This causes not only a bad back,

BUTT ——!

Finish Dates

> **DON'T** ever sign a contract that does not have a finish date (you can be flexible within two weeks).

Remember I mentioned the project taking six years instead of six months to finish? This again was one of the horror stories I spoke of earlier.

Not having a finish date written into the contract is like not having a finish date for a pregnancy. Would you like both to last a year?

Chapter 11

It's a Partnership

Let's make this a team effort. I remember when I was just a kid learning about the dynamics of a successful football team. A good team is the sum of its parts:

- The owner (you).
- The coach (the contractor).
- The players (the subcontractors).
- The Super Bowl (your finished project).

All possible because of your due diligence as the owner—putting together a championship team, with an outstanding coach. A coach that you hired. **GOOD JOB!**

During your remodel, it may seem that everything in your life that goes wrong is the contractor's fault.

- The kids are flunking out of school.
- My husband isn't speaking to me.
- The house is a mess.
- The driveway is always blocked.
- We are getting lung cancer from the dust.
- The dog wants to go home with you now.
- *We want to take the dog home with us now.*

Here is a story about a disgruntled customer with a complaint about her kitchen project. I hope you not only enjoy reading about it, but that it makes you think twice before pointing the finger of blame at someone (and I don't mean the middle finger).

I had just finished a kitchen for a single mom, a really bright mathematician, working two jobs at high tech companies, and raising two teenage sons. I don't know from personal experience what it is like to raise two kids on your own, but from my observations here, she had her hands full.

It was a little stressful trying to work around the kids' schedules and hers. The kids did not like the disruption, and she didn't like having the house torn up and creating more stress on top of her normal life.

I remember in the initial stages of the project the lady asked me for a suggestion on which stove is the easiest to clean and take care of. I suggested the new glass smooth top cooking surface. No burners to spill things into, and you simply wipe the surface clean.

I got a call about two days after the completion of the job from Ms. X. She was irate!

"YOU GET YOUR BUTT," (she didn't say butt), **"OVER HERE TO MY HOUSE."**

"I CAN'T COOK CORRECTLY ON THIS NEW SURFACE."

"I HAVE TWO KIDS TO FEED."

"I BURNED THE EGGS, THE PANCAKES, THE BACON."

"I CAN'T CONTROL THE HEAT."

"YOU TOLD ME TO BUY IT, YOU TOLD ME IT WAS EASY, YOU TOLD ME, YOU TOLD ME, YOU TOLD ME, YOU TOLD ME."

If this poor woman had ten fingers on both her hands, she used eleven somehow that morning.

I told her to relax, maybe I hooked it up wrong, and I'd be right over. Hooking up a black wire to a black wire, a red one to a red one, and a white one to a white one doesn't take a Ph.D But, she had a Ph.D and I didn't, so off I went on a Sunday morning.

She met me at the front door very upset. **"LOOK AT THAT THING. I BURN EVERYTHING. THE GLASS COOKING SURFACE IS POOP."** (She didn't say poop.)

Well, I went in, looked at her son with the *attitude* tapping his pencil on the counter thinking, "My mom has you now Mr. Contractor."

Next, I saw that she was indeed telling me the truth. Everything was burned.

Remember I recommended the glass cook top? The one with the smooth glass cooking surface? Well, I guess some of us can't see on either side of a straight line.

SHE WAS COOKING ON THE GLASS! DIRECTLY ON THE GLASS! She wasn't using any cookware!

I'M GLAD SHE WAS A PH.D AND NOT MY MD.

I am not a very religious man. I don't necessarily believe in miracles, but it was truly a miracle that I could keep it together without bursting out in laughter.

Just when I figured out what to say to her without embarrassing her in front of her son, she blurted out, **"AND IT ONLY HAPPENS IN FOUR SPOTS!"**

DUH!

I suggested that we get the instructions. I told her that with some of these new contraptions you can only use steel pots and pans, not aluminum or copper ones. Perhaps she was using the wrong cookware. I glanced over at her son. He looked down shaking his head. **HE GOT IT.** Sorry Ma'am. The lady looked me right in the eye and said she would have to find the owner's manual. She would get back to me.

THINK SHE EVER DID?

THINK I USED HER AS A REFERENCE?

THINK SHE WENT TO COSTCO?

THINK HER KIDS WILL BRING UP THE UNFORTUNATE EXPERIENCE AT EVERY FAMILY FUNCTION FOR THE REST OF THEIR LIVES?

THINK I RECOMMEND THOSE STOVES ANYMORE?

Remember, be nice. We try our best. This is a partnership between us with a common goal.

Chapter 12

In Closing

Let me briefly summarize what we have covered in this book. Although you are most likely brilliant now, I feel that this may be necessary just as a refresher.

1. How to choose a contractor, designer, salesperson
Ask the right questions, don't be shy.

2. Who to throw out of your house, and when
The little red flags are waving in front of your nose. Now you know how to recognize them.

3. How to check out sub contractors
Use the references.
Call the Contractor State License Board.

4. How to check out the licenses
The Contractor State License Board.
On his truck.
License classifications.

5. The CORRECT way to use references
Ask them anything.
Cross reference the subs.

6. How to get proof of insurance
Certificates of Insurance naming you as an additional insured.

7. How and why to go through the permit process

8. What permits are needed and why
Plumbing
Electrical
Building
Mechanical

9. How to avoid winding up with a lien on your house
Unconditional waivers of lien.

10. Controlling your money, saving your money

Payment schedules.
A rather large balance at end of the project.
Fixed prices in the contract.
Work change orders.

Although much more was covered, this should summarize what I wanted to **HAMMER INTO** your brain.

I hope this book helps you with all of your remodeling projects for the rest of your lives. You really can apply it to just about any project, from something as small as painting, to remodeling your entire home.

ALL OF THE EXAMPLES AND STORIES ARE TRUE.
ALL HAVE HAPPENED DIRECTLY TO ME. Of course I have **SPICED** them up a little, but this is about kitchens, isn't it?

I know I presented information in a very negative way and most likely have stereotyped salespeople, designers, and contractors into a very special category of subspecies. In reality this in not the case. Most of us are hard working men and women doing our best to help create what we feel are incredible living and cooking spaces.

Just a few **BAD APPLES** ruin it for the entire **BARREL**. Since you have a different career than we do, it limits your ability to be **A GOOD PRODUCE MANAGER** who knows just exactly which apples to throw out.

Maybe now you do.

Using every step in this guide can at times be mentally draining and seem like a lot of work. Remember, you will probably do a project like this only once or twice in your lifetime.

Don't make it a **DRESS REHEARSAL.** It isn't difficult to make the **MAIN PERFORMANCE A MASTERPIECE.**

Thank you,

Steven Katkowsky
Steven's Incredible Kitchen Machine

California License number 457095. *Better check it out!*

www.stevensincrediblekitchenmachine.com

p.s. Look for our next book to be out soon. The content will be the job process—from deposit check to balance check. I plan to document both a complete kitchen and bathroom remodel from start to finish.

A great **how to** guide helping you to know when your contractor is falling behind or mismanaging your project.

ANY SUGGESTIONS FOR A TITLE?

www.ingramcontent.com/pod-product-compliance
Lightning Source LLC
Chambersburg PA
CBHW022058170526
45157CB00004B/1393